What <u>you</u> can do to reduce climate change

Seth Wynes

EBURY
PRESS

Contents

Foreword

Nothing about climate change is easy. But that's where we are, right now, at the most important moment in human history. Our planet, the only place in the known universe capable of harbouring life, is in a crisis of our making. If we don't act immediately, together, with everything we've got, we have a good chance of losing everything.

That's the difficult, terrifying reality of our time.

We've reached a moment in our multi-million year history as humans where every action and every life in every corner of our world are intimately connected. In miniscule and enormous ways, humanity's decisions have altered our collective fates, and we're now at a crisis point.

We've got to realize that these actions – our actions – matter. Not only do they matter, but when we act as though they do it reminds us of the biggest truth about the biggest problem facing humanity: we're all in this together.

The sheer scale of what we're facing is enough to cause paralysis in most people. I know that and it took a long time for me to admit to myself that not only are things as bad as scientists say, but that, starting with my own actions, I have a duty to help change the narrative. But by reading this book, you're showing that you're ready to be a leader at exactly the right time, when it matters the most.

This book brings something to the conversation that has been missing for a long, long time: an indispensable guide to which actions are most important at this most important moment.

When I first read Seth Wynes and Kim Nicholas' research results showing what types of individual climate action make the biggest difference, I remember breathing a literal sigh of relief. In my years as a climate journalist, by far the most important question I get from readers is, 'What can I do?'. Seth and Kim had answered that question.

When coupled with collective action efforts, like voting, strikes and non-violent protests, the steps outlined in this book, scaled up over millions of people, bring the transformative change that we so desperately need.

For too long, the conversation around action on climate change has been framed around sacrifice and blame. At last, the conversation is changing to be one that is more inclusive, more just and more positive. What's more, scientists tell us that, building a future that preserves human civilization in our lifetimes is only possible through 'transformative' action. A liveable future is only possible through a radical re-imagining of how each of us live.

Not only is our situation not hopeless, but it *can't* be hopeless. Talking about climate change with friends and family members is the first step to imagining how we can make the world a better place together. By putting our words into action, we'll become living examples of the kind of world that simply must exist.

Thank you for opening this book. I know it's not easy to embark on the journey you're about to take. But I do know that nothing could be more important.

Eric Holthaus

Introduction

No matter where you are, there are likely changes to our climate happening now that already affect you. Maybe it's longer dry spells and a new risk of heatstroke. Or the arrival of new pests, like ticks or mosquitoes, in your region. It could be something relatively minor like a shorter ski season that has you worried about the future of your favourite hobby. Or perhaps something major, like more frequent wildfires, longer droughts or even more intense tropical storms crashing on your coast. If you're reading this book I'm guessing you're already somewhat aware of these things. But why should **you** in particular do something about them?

Why You Matter

If your actions are only a drop in the ocean, why bother trying? Let's start by looking at those drops that we're contributing. In this case we're not worried about water, but carbon dioxide and other greenhouse gases (like methane) that are causing the planet to warm. And every kilogram of carbon dioxide matters quite a bit. Unlike an orange peel thrown out of a car window, carbon dioxide lasts for a long time – for centuries or even millennia. One kilogram of carbon dioxide (about the amount that would be generated from driving 7km in a mid-size vehicle) can melt 650kg of glacial ice over its long, long lifetime.[1] Your actions leave a legacy that is recorded in our atmosphere, melts our glaciers and is measured by our rising oceans.

"But there are a lot of other cars on the road," you might say, "aren't individual actions a little limited?" And you're right. That's why this book has a whole chapter on collective action (see page 69). But even so, individual actions that reduce your carbon footprint are important, because they don't occur in isolation. You probably don't eat every meal alone, or go on vacation without posting photos, or buy a new vehicle without your neighbours noticing. And the science says that this matters:

- Homeowners are more likely to install solar panels when they see their neighbours leading the way.[2]
- Cafeteria goers were twice as likely to select a meatless meal when they were told that more and more members of the public are limiting their meat consumption.[3]

> Households decrease their energy use when told that their neighbours use less energy than they do.[4]

It's a very human response to follow the crowd, and that can have positive consequences if the crowd is shifting towards sustainable behaviours. When a crowd becomes a movement, governments have to start listening too. Cultural changes can happen faster than you might think, but our society needs leadership from individuals like you to get there. In this book I'll show some of the most important areas where we need that kind of everyday leadership.

Make Your Actions Count

In 2017 I published a scientific paper that identified the most effective things we can do about climate change and investigated whether governments communicate those actions to the public. I found that for the most part, government documents and high-school science textbooks focused on moderate or low-impact actions (things like turning off your lightbulbs), while neglecting the things that can make a much bigger difference. Now a lot of those smaller actions are great to do, but people have known that they should be recycling for a long time, and climate change is such a big threat that we need to be tackling it with the kinds of actions that are on the same scale as the problem.

A recent report from the Intergovernmental Panel on Climate Change (the ultimate authority on the subject) outlined how the decisions we make in the next decade will largely determine whether we maintain a liveable atmosphere on our planet or instead, imperil our own civilization. In this book I'm going to focus on the type of behaviours that can have a big impact in that time frame: how we travel, what we eat and the collective actions that can drive political and societal change.

As we discuss the actions that can have the greatest impact on your carbon footprint, different behaviours will matter more for different people. If you're the CEO of a large corporation you can probably reduce greenhouse gas emissions more through your company than you can through changing your diet. But this book is aimed at the average person living in a democratic country.

Because we're focusing on personal behaviours, I'm drawing from research that also focuses on the individual scale. By doing so, I will avoid the confusion that sometimes comes from taking global numbers and trying to apply them to our own lives.

As a quick example, agriculture is responsible for more emissions at the global level than aviation (about one fifth of all global emissions come from agriculture,[5] compared to about one twentieth for aviation[6]). Someone can look at those numbers and say that obviously it's more important for the average person to focus on what they eat rather than how often they fly. But everyone eats food and only some people fly in planes. If you're among the fraction of people who do fly on a regular basis, your footprint from flying is going to be larger than your footprint from food,[7] and so you need to consider the impact your travelling AND eating habits are having on the planet. That's why throughout this book I'll not only be identifying the actions you can take and the impact they will have on your carbon footprint, but also debunking some of the myths that surround these actions.

Measuring Your Impact

The best way to compare different actions is to use the same measuring stick. For climate change, that measuring stick is how much greenhouse gases are reduced by engaging in an action. We can measure greenhouse gases in tonnes of carbon dioxide equivalents, or tCO_2e. To give you an idea of how big these values are, the table on page 14 shows a cross section of countries with some of the highest emissions per capita and the lowest, with a focus on Europe. In the UK each person produces about 7.4 tCO_2e per year on average.[8] This is lower than a few places (like the United States and Australia) but much higher than many others (most developing nations).

There are many reasons for the differences in per capita emissions between countries. Colder countries, like Russia, need more winter heating. People living in larger countries or those that are in more sparsely populated areas, like Canada, with less public transport, tend to drive more. Countries where the electricity grid is powered by coal, like Australia, generate a lot of emissions through their energy consumption.

We can't make excuses for our own carbon footprints by looking at the emissions of others. For example, some might look at India's huge emissions and think that climate change is their problem. But if we look at the data, we can see that individuals in the Western world have a much larger footprint per capita than people in India.

You could also look at per capita emissions of a country that is not your own and find them depressingly high compared to your own country – what difference can you make, when they can make so much more of a difference? The way I look at it is that if you were on a sinking boat you wouldn't stop bailing water just because the person next to you has a bigger bucket to work with. We all need to contribute as much as we can. We have a collective responsibility.

Nation	Per capita greenhouse gas emissions (tCO_2e)[9]
Australia	22.8
United States	20.2
Canada	19.4
Russia	18.0
New Zealand	16.9
The Netherlands	11.5
Germany	11.1
Finland	10.7
Japan	10.2
Denmark	9.0
United Kingdom	7.4
Italy	7.2
France	7.2
Spain	7.0
Sweden	5.4
Peru	2.8
India	1.8

A quick disclaimer that these numbers are complicated. For instance, if you buy a German car, or a sweater made in India, you are responsible for emissions that are going down on someone else's ledger.

If you don't live in a country at the bottom of the per capita table, your emissions are higher than what is sustainable. By the year 2050 the global average for per capita emissions needs to be at 2.1 tCO_2e per person per year if we are to keep warming below a level that the international community has agreed upon as an upper limit (a 2°C increase in global temperatures).[10]

No amount of warming is really safe, so we want to avoid any increase that we can. That's part of why every action matters. Even so, this at least gives us a kind of target that we can aim for. We'll keep that number in mind as we examine some of the best actions you can take for the climate.

1

Getting
Around

One of the great advantages to fossil fuels is that you can take an energy-dense substance and burn it whenever you need to move people around. Because of this, we've created a mobility system where fossil fuels dominate and, unfortunately, that makes transportation a big driver of climate change. If we want to fix the problem of our world becoming too hot for us to live in, then we need to change how we get around. Let's start with a lighter topic – vacations, then look at the everyday changes we need to make.

Holidays

The lowest-carbon way to go on holiday
is to keep it local. You can be an eco-saint
once you arrive at your destination – reuse
your hotel towels, eat local cuisine, nurse an
injured bonobo back to health – but if you
traversed an ocean for ten days of fun in the
sun, all of this will be eclipsed by your travel
there. Staycations have actually become
popular without climate change being a
motivating factor – you can skip the hassle of
booking flights, getting travel insurance and
exchanging currency, and save a lot of money.

Travelling to a local destination without
burning a ton of carbon is possible if you
move over ground to your destination. Trains
are best, coaches and buses are good as well,
cars are not great but are better than a plane

(partly because you're not as likely to travel outrageous distances).

Transport method	KgCO$_2$e to move one passenger one Kilometre[1]
European high-speed electric rail	0.0302
Coach bus	0.0355
European diesel rail	0.0657
Ferry	0.1378
Car*	0.146
Plane (travelling within Europe or an equivalent distance)	0.2135

*Based on a typical car with an average UK value of 1.57 passengers per vehicle.

If you have high-speed rail where you live, it's possible to arrive at your destination faster than if you took a plane, which is a huge plus. Other perks include: no queue for security and no ear-popping when you approach your destination; you get to stretch your legs on the walk to the dining car and enjoy a sense of connectivity to the landscape as you watch the scenery change around you.

The holidays that I've taken further afield by rail have involved stops in multiple cities, seeing different people and sites along the way. When you fly for vacation the travel is something to get over as quickly as possible. But when you travel by train the journey can be part of the holiday.

In spirit, a cruise might seem similar to a train journey. While they may share the same form of having multiple stops, they are not similar from an environmental view. If you're looking for a relaxing way to take time off work but wanting to help reach our carbon target, then a cruise is not for you. The average emissions of a cruise ship passenger range from 1.56 tCO_2 per trip, up to 6.3 tCO_2 for Antarctic cruises.[2] In the case of the Antarctic cruise, the emissions for that single trip are larger than what the average world citizen produces in a year. Unfortunately, that number doesn't even include the emissions from the flights that many people took to get to their port of departure. And flying, as we'll see, is quite the climate culprit.

Air travel is a huge problem for our planet. While aviation is responsible for a relatively small fraction of global emissions at the moment, the number of flyers around the world is rapidly increasing, and by 2050 this sector could take up 22 per cent of global CO_2 emissions.[3] Every year airlines make improvements in efficiency, but this just isn't enough to cover the growth in the number of passengers.

There are no easy techno-fixes for air travel.

Every so often you may see a headline about an exciting development in aircraft technology: new biofuels that will make planes environmentally friendly; electric planes that are close to becoming mainstream. But the unfortunate truth is that none of these solutions are taking off fast enough.

Electric aircraft are one obvious solution to the sustainability issues that plague aviation. Technology will decarbonize personal vehicles mostly by using batteries, so why not do the same for planes? The answer is that batteries are heavy, and planes are hard to lift. The further you want to travel in a plane, the more energy you need, and the bigger and heavier the battery will have to be. It's not impossible, but this technology is more tailored for short-distance travel (the kind of travel we should be replacing with energy-efficient, high-speed rail).

Researchers think that electric aircraft might be well-suited to a range of about 1100km, but they don't anticipate them being able to operate more than 2200km within the coming decades.[4] That means a trip from London to Berlin would be feasible in an electric aircraft (966km), whereas a flight from London to New York (5555km) would be out of the question.

Biofuels are theoretically a climate-friendly alternative to jet fuels. The idea is that if you take a plant, convert it to fuel and then burn it, the whole process is carbon-neutral so long as you put another plant in the ground to start the cycle over again. But there are bigger problems afoot. Farming biofuels requires space, some of which might be high-quality farmland. As we try to feed a growing planet, should we use land that could be used for food to grow jet fuel instead?

We could use crops that are inedible and can be grown on unproductive soil, like the jatropha plant or algae, but estimates from the ICAO suggest that we would need an area one third the size of Australia to power global aviation with jatropha and land the size of Ireland to do so with algae.[5] Ideally we should be thinking of preserving some of those areas for the sake of the natural world – even at the "upper limit" of 2 degrees of warming that we are already approaching, one in twenty of the globe's species will be threatened with extinction.[6] We don't want to add habitat loss on the scale of small nations to the obstacles those species have to overcome.

Most experts who study climate and air travel have looked at these numbers and concluded that society needs to reduce demand for air travel if we don't want to overheat the planet.[7] Reducing demand is a fancy way of saying flying less. And that can be a hard thing to swallow in our society.

Flying is linked with status in our culture. If your company pays for you to go to conferences or meetings abroad, we view this as a sign of success. Younger people consider trips to other continents as a rite of passage, and the accumulation of passport stamps as social credit. Browse through millennial online dating profiles and it shouldn't take you long to come across phrases like "World traveller, 54 countries and counting". When describing what we view as the most attractive parts of ourselves, nobody would write, "I leave my SUV idling in the driveway and litter in national parks", but people advertise having a gigantic carbon footprint when it comes from frequent flying.

Air travel has accomplished great things in our world, but it's time we started viewing it as a precious and limited resource. You can be a part of change not just by avoiding air travel yourself, but by helping to change the culture.

- Push for your workplace to allow extra time for business trips so that you can take a slower form of travel.
- Write to newspaper editors and ask them to consider not accepting ad money for international travel destinations.

- Use videoconferencing and teleconferencing at work and encourage your company to upgrade to the latest technology (it's a good business decision since travel is so expensive compared to one-time technological investments).

Myth: The plane will still leave without me

This type of thinking applies to a lot of environmental actions. "But if I don't buy the burger, someone else will buy it." Companies plan how much of a product to offer by using sales as a feedback mechanism. So every time you buy a plane ticket you are sending a "market signal" – instructions that you want more flights to take off and more runways to be constructed. While *this* flight will still take off without you, the next one will not if enough of us make the same decision. And preliminary evidence suggests that people are more likely to reduce their flying if they know someone else who has cut back for climate reasons[8] – so there is value in your leadership.

Protesting against airport runway expansion or supporting politicians who are willing to implement a frequent flyer levy would go a long way to curbing the emissions in this runaway sector – check out the Collective Action chapter on page 69.

Do you have to fly?

Each flight that you skip drastically cuts your carbon footprint. Here's how much you would save by skipping out on a few round-trip, economy-class flights:

- London to New York: $1.8tCO_2e$
- New York to Mexico City: $1.1tCO_2e$
- Sydney to Tokyo: $2.5tCO_2e$
- Berlin to Bangkok: $2.8tCO_2e$

You might be able to skip some of these by alternating your annual holiday so that one year you do a staycation. You could also plan trips over ground by train instead!

Everyday Travel

Living car-free is one of the most effective
things you can do for the climate. My research
found that this single action could reduce the
average person's carbon footprint by 2.4 tCO_2e
each year. The surface-level reasons for this
are obvious: when you drive a personal vehicle
you put a fossil fuel in the gas tank, burn it and
pump out planet-warming greenhouse gases.

Electric vehicles are much, much better for
the climate, but they're still personal vehicles,
which means they're still problematic.
Whether you're in an electric vehicle or a
hybrid or an internal-combustion vehicle,
fossil fuels were used at various stages of its
construction (carbon emissions were released
to mine the metals, process those metals,
whether they are steel or aluminium, make the

battery, and run the factory where the car was assembled). The car has a big carbon footprint before it drives off the lot.

Hopefully these processes will be decarbonized soon, and maybe we'll do more recycling and less mining in the future, but until then personal vehicles carry an invisible cost.

Cars

Personal vehicles like cars force us to build our cities in inefficient, inherently polluting ways. First you need to create roads, petrol stations, overpasses and parking spaces. Constructing this infrastructure, of course, creates emissions (and much more than the emissions needed to make bike paths and bike racks). And this construction spreads the city apart. Every three-lane motorway, every car park and every parking space is extra distance that you have to cross to get to work or get to the park or get anywhere. Sprawl makes it harder to move across a city in a car or bus, but especially hard if you cycle or walk. Sprawl also demands more car ownership to shorten those long distances, which creates the need for more parking, and soon there's a feedback loop that results in an unwalkable city.

Giving up your car is the absolute ideal. It is a great decision for the environment and there are other health benefits:

- One study found that each additional hour spent in a car per day was associated with a 6 per cent increase in the likelihood of obesity.[9]
- Air pollution caused by vehicles shortens lives by an average of 5.4 months.[10]

- You're about 66 times more likely to die driving the same distance in a car as taking a bus.[11]
- Normally longer commute times are bad for your mood – but active commuting (walking and cycling) can be so good for your mood that increased *active* commuting can actually improve your mood![12]

By living car-free you are opting out of a system that is harmful to yourself and others. Thankfully some recent societal changes and new technologies are making it a lot easier to go without a car.

There are electric bicycles, electric scooters, scooter-sharing services, bike-sharing services and apps that let you know the fastest route by public transport and how many minutes away your next bus is from the station. While all cars are inherently problematic (see page 33), electric bikes and scooters let you move quickly through the city with a much smaller carbon footprint (both in terms of emissions and from constructing them).

Motorbike users can reduce their burden on the environment by switching to electric. Rapid technological improvement in electric motorbikes will help drive down costs and reduce charging times.

If you can move from driving everywhere in a car even just to a mix of options then that is much better than the status quo. Where I live in Vancouver, I'm fortunate enough to have a lot of options for getting around. I can walk for local errands, commute with public transport, and if I want to go hiking well outside of the city then I can use a car-share service.

What if you can't give up your car?

There are situations where it's difficult to not use a car. In those cases, you should go electric.

Electric vehicles require less maintenance, have lower fuel costs (and your government may offer rebates for purchases of the vehicle or the charging station, depending on where you live), run quietly and emit no local emissions. My research found that on average, switching from a petrol car to an electric car saves 1.25 tCO_2e per year, but the savings keep getting better as our electricity grids make use of more and more renewable energy.

There used to be a worry that an electric vehicle would run out of charge halfway to a destination. Fortunately, as battery technology

improves, so does the range they can handle on a single charge (Tesla advertises a range approaching 500km for the Model S). And as these vehicles become more popular, so will the charging infrastructure needed to support them. For now, going electric is a choice, but based on trends in legislation we can tell that it won't always be:

- At least half of all new vehicles by 2030 must be ultra-low-carbon in the UK.[13]
- All cars sold after 2040 in British Columbia will be Zero Emission Vehicles.[14]
- 50 per cent of new cars in Norway are **already** hybrid or electric vehicles.[15] Note that contrary to some beliefs, electric cars do still work in the cold – Norway isn't exactly tropical.

Electric vehicles are inevitable for those who need to keep driving; why not join the winning team early?

Some people face legitimate structural barriers in going car-free. They might be living in disconnected suburbia, or low-density countryside, or just a city that is built for cars and not humans. In that case, I really recommend pushing for changes at the city/ municipal level.

Let's start with a single example that sounds a little nerdy, but has huge implications: road connectivity.[16]

A city that's laid out in a grid has high connectivity, and a city with a lot of culs de sac or three-way intersections has low connectivity. High-connectivity cities are more likely to have grocery stores and coffee shops in every neighbourhood, because you don't have to walk so far to reach them.

Low-connectivity areas are hard to serve with public transport because neither buses nor the people trying to get to their stops can travel in straight lines. Instead people stay in their cars and drive more than they would in a connected city. All of this is a recipe for more greenhouse gas emissions.

Researchers will tell you that roads are incredibly static; if you build a house it might get replaced by a block of flats or a café, but the layout of a road network is likely to remain the same, even if half the city burns down.[16] That means city councillors and urban planners are making decisions today that will have impacts far into the future.

You can get involved in issues like this when new land developments are proposed or when new zoning bylaws are brought forward. Cities often hold public consultation periods. In that case, having your say can mean showing up to a townhall-type meeting or submitting comments.

There are also other, more visible changes that you can fight for. Greater funding for public transport, more protected bike lanes, congestion charges for central city drivers and pedestrian zones are all good bets for making car-free living easier, safer and more fun. Voting for political representatives who support car-free mobility is one important step in moving towards a sustainable future (see page 75 for more on why your vote matters).

You might hear that there are greater emissions from the extra food needed to supply the calories for a bike ride. But few people in Western nations are calorie deficient;

we already get more calories than we need. Cars are an extremely inefficient way to move people around because only a tiny portion of a vehicle's weight (about 5 per cent) comes from the passengers. A car burns most of its fuel transporting itself, not you.

Tips for new cyclists

If you're switching from a car to a bike you're going to save a lot of money. Think about reinvesting at least some of it in the bike to make your trips more worthwhile! Get a proper helmet, a bell, a lock, both front and back lights and a comfortable saddle. While you're on the bike, here are some tips:

- Ride about one metre from the edge of the road so that drivers are more aware of you and give you a wide margin when passing

(your safety is more important than their convenience). This makes you ride in a predictable straight line instead of dodging gutters and sewer grates.

- Roll your right trouser leg up to keep it free from moving parts, if your bike doesn't have a chainguard.

- Plan a safe (and lazy) bike route. Sometimes the more direct way isn't the fastest, and it's better to go a little out of your way to stay on a bike path.

- Slow down and communicate with pedestrians on multi-use trails. Ringing your bell when you're approaching or calling out "on your left/right" to pass those ahead of you is the respectful thing to do.

- Take up the whole lane when stopped at a red light to prevent cars from dangerously squeezing in beside you.

2

What We Eat

Humans have unrecognizably transformed the landscape in order to feed ourselves. Of all the world's land that is not covered by desert or ice, 43 per cent is occupied by agriculture.[1] And roughly one quarter of all human-made carbon emissions are created by the food supply chain.[1]

Governments and engineers can address some aspects of the climate problem with legislation or technological changes, but the experts on land and food say that we need to change the way we eat if we're going to avoid dangerous levels of climate change.[2]

Starting with the changes that mean the most, the best things you can do to reduce the effect of your diet on the climate are:

1 Eat an all-plant diet – vegan diets reduce emissions slightly more (about 0.9 tCO_2e per year) than vegetarian diets (about

0.8 tCO$_2$e found in many studies)[3] largely because they don't include dairy products, but both are great.

 Eat fish and plants instead of poultry and red meat (about 0.85 tCO$_2$e reduction per year found in one study) – a pescatarian diet has a lower carbon footprint than a standard Western omnivorous diet,[4] though there are ecological concerns with eating a lot of fish.

 Eat pork, poultry and plants instead of red meat (beef and lamb) (about 0.56 tCO$_2$e per year).[5]

 Eliminate food waste (0.37 tCO$_2$e per year).[5]

⑤ Choose seasonal/somewhat local food over anything that's been transported by air, or has been grown in greenhouses, as both have a high carbon footprint (0.16 tCO$_2$e per year).[5]

Why Is Meat Bad For The Climate?

As the global population increases to a projected 10 billion people by 2050,[6] it will become critical to find ways to make sure everyone can eat. One group of researchers calculated that if food was only grown for us to eat (not to produce biofuels or to feed livestock) we could feed an additional 4 billion people.[7]

So how does meat come into this?

Land use

Imagine you have two plants, both identical in size, nutrients and calories. You feed one of the plants to a cow and you eat the other one. The first plant is digested by the cow, and the cow uses the nutrients and energy from that plant to walk around, breathe, digest other food and to add on a little more muscle. Later you slaughter the cow, eat it and get back some of the calories that you added with the plant. But not all the calories. With the plant you ate, although you are not 100 per cent efficient at digesting it either, you get a lot more calories out of this second plant then you do out of the one that took a cow detour.

This is one issue with eating meat instead of plants – it is a less efficient method of growing food. We can put numbers to this phenomenon: if you took the same cropland used to feed egg-laying hens but switched it to a nutritionally similar plant-based food (let's say tofu), you would produce twice as much food from the plants as you would get from the eggs. And if you did the same thing with cropland used to feed beef you could get

20 times as much plant-based substitute![8]
This need for more land means more fertilizers
(which cause algae to bloom in the oceans and
kill off fish), more deforestation and much
more greenhouse gases.

There are examples of farming sheep, goats
and cattle that produce lower carbon emissions:
just as we can grow biofuels on land that's no
good for edible foods, we can also raise livestock
on land that is too rocky or arid for growing
crops that people could eat. But one study in
Sweden found that only farming livestock with
sustainable grazing, and feed from food waste,
would require a 60 to 80 per cent reduction in
meat from the average Swedish diet.[9]

It is impossible to meet our climate
targets without making a shift in our diet away
from meat.

Greenhouse gases

Beef production in particular is problematic for how much greenhouse gases it generates, because cows happen to produce a lot of methane, a greenhouse gas that is a powerful, short-term warming agent – about 28 times more potent at trapping heat than carbon dioxide on a gram for gram basis.[10] Beef and lamb are both problematic, but as the world, in general, eats much more beef than lamb, beef is a more critical problem for us to tackle.

Dairy is not as intensively produced as red meat, but it is still not the best climate option. Per gram of protein, cheese results in 5.5 times as much greenhouse gases as tofu, while cow milk results in 3.2 times as much greenhouse gases as soy milk per litre.[1]

Researchers have made efforts to reduce the amount of methane that cows produce by using feed additives (including seaweed), and it's great that options like this are being investigated. But these technologies are not yet ready to be scaled up,[11] and do not solve the growing issue of space.

Resource intensity

The intensive farming methods, especially those required to keep up with demand for meat, need nitrogen fertilizer made out of natural gas and the mining of phosphorus to fertilize crops, and the burning of fossil fuels to operate farm machinery, and deforestation to clear land for grazing.

The Western world eats the most animal products. In 2009, data showed that the richest 15 nations (Australia, Austria, Canada, Denmark, Finland, France, Germany, Ireland, Japan, the Netherlands, Norway,

Sweden, Switzerland, the United Kingdom
and the United States) ate 750 per cent more
meat protein per person than the 24 poorest
nations.[12] As nations become wealthier, their
citizens eat more meat. If you combine that
with a growing global population, you run into a
difficult situation. How are we going to feed the
world if everyone wants to keep eating meat?

We can't farm enough sustainable meat to
satisfy the world's demand.[13]

Beans for beef?

Probably the most common question I get asked
is: these individual actions are nice, but do you
actually believe they can make a difference?
My favourite study that describes how a small
change in how we live could achieve big results
was written by some researchers asking a
simple question: what if everyone in America

swapped beef out of their diet and replaced it with beans?[14]

This one change would free up 42 per cent of US cropland and achieve about half of the greenhouse gas reductions that America needs to reach its 2020 target (which it will miss). One change with a huge difference.

How can you start eating less meat?

There is a lot of misinformation surrounding meat. Perhaps the biggest misconceptions involve the idea that you can't lead a healthy life without eating it. The fact that vegetarians live at least as long as omnivores, or possibly longer,[12] should sort of dispel that notion. But are vegetarians and vegans getting enough protein? The fact is, most people on a Western diet get more protein than they need. The average person in America, for instance, consumes 36kg more lean meat per year than the recommended upper limit.[15] Here are some high-performance athletes who thrive on plant-based diets:

- Patrick Baboumian (Germany's strongest man in 2011)
- Venus Williams (professional tennis star)
- Alex Honnold (climbed the mountain El Capitan without a rope)

It's true that not all plants are sources of complete protein (meaning they don't contain all of the amino acids that your body needs). But the US Academy of Nutrition and Dietetics says that vegetarians who eat a **variety** of plant foods and meet their daily caloric needs get enough of each type of amino acid to be healthy.[16]

Perhaps you live somewhere where it's not easy to choose vegan options to cook at home or eat out? Even starting small is at least starting somewhere:

- Pick one day of the week to go without meat.
- For meals where you do eat meat, don't make it the focus of your meal. It should occupy a smaller portion of your plate but ideally be left out completely.
- Try meat substitutes. There is a huge range of plant-based alternatives for everything from burgers to bacon, and some of them are remarkably good imitations. Just be aware that health experts recommend limiting your intake of processed foods.
- Replace dairy milk with a plant-based milk (all plant milks have a substantially lower climate impact than cow's milk).

You might have heard that soy and nut farming is bad for the planet and so you should not choose these options. At the moment, over 70 per cent of American-grown soy is used for animal feed.[17] From the perspective of the climate, drinking one glass of cow milk is the equivalent of drinking one glass of soy milk and then pouring two glasses of soy milk down the drain. Different non-dairy milks each have their own nutrient profile and require various amounts of land, energy and water to produce. But because they are at the bottom of the food chain, they require less of these inputs than dairy milk, and produce fewer greenhouse gases.

Eating a plant-based diet (or plant-rich diet) comes with a lot of benefits not related to the climate. We've already talked about the need to feed the planet, and the need to conserve land for biodiversity. But even if climate change wasn't a problem, and if biodiversity loss wasn't happening, eating more plants has been associated with:

- lower rates of cancer, coronary heart disease and Type II diabetes[12]
- lower BMI (body mass index); high BMI is a risk factor for mortality, even when researchers control for smoking and physical activity[18]
- improvements in mood[19] (the evidence on this is still preliminary, but reasonable).

Beans, as a specific example, are extremely healthy for you. They're high in zinc, iron, magnesium, folate, fibre and protein. They are associated with lowered risk of diabetes and heart disease and are also extremely cheap.[20]

Swapping from beef to beans is the kind of positive change you can make on your own or with your family. But you can also make changes in the community around you.

- Own a restaurant? Manage a canteen? Provide a plant-rich menu for your patrons. Have your menu default to vegetarian (customers ask or pay extra for meat in their salad, rather than making them ask for the dish without chicken).
- Ordering food for your department's meeting? Pick a veggie option (depending on your workplace culture it might take some effort to get buy-in).
- As of the writing of this book there isn't a lot of political will to disincentivize meat consumption, but that is likely to change. As it does it will be important that everyday people support elected officials who are willing to make bold proposals like taxing foods based on their carbon footprint.

Food Waste

If we waste food, then we have to replace those calories that are spoiling in a landfill site or our compost bins, which means buying more food and putting more pressure on the environment. A lot of food waste occurs before the food arrives at a grocery store, and is somewhat outside of our control (e.g. the food that spoils on a lorry between the farm and the market). But the food that goes into our bins is something we can address. Some ideas[21] to help you step up your game here:

- Plan meals in advance.
- Store apples and carrots in the fridge.
- Label your food with the date so you're reminded to use it in time.

- Keep perishable fruits in a visible place (bowl on the counter) to encourage you to eat them in time.
- Don't give up on wilted vegetables. Broccoli and celery can look sad and limp but will revive themselves if placed in cold water – there's a reason grocery stores mist their veg sections!

Airmiles and Greenhouses

Avoiding food transported by air or grown
in commercial greenhouses/hothouses is
beneficial, but these are fringe actions in terms
of effectiveness. Still, if you've addressed
everything else in this chapter it's worth
paying attention to. Food grown in hothouses
requires a lot of energy to keep warm in winter
months, and food that's transported by air can
have a large carbon footprint. Food can come
from far away, but if it's travelled on a freight
ship the carbon footprint is actually not that
bad. Seeing a food label with a foreign country
doesn't necessarily mean that your food has
been on an plane. It might surprise you, but
data shows it is often better to eat your food

from another country if that country grows it more efficiently, just so long as it came by boat or rail and not plane. If in doubt about the transportation of your food, advice that applies to both airfreight and hothousing is to eat produce that is in season and locally grown.

Food Miles

What you eat (plants vs meat) matters a lot more to the climate than where it was grown. Now there are a lot of good reasons to eat local: it's good to support your local community, and to know how the food that you are eating is grown. But it won't make much difference to the climate. As I mentioned above, food that's been transported by air has a huge footprint, but food can be transported great distances by boat without generating a big climate impact.

One study found that in the United States, 83 per cent of a food's carbon footprint comes from production, and only 11 per cent from transportation.[22] This means that if you're only worrying about transportation, it's pretty easy to make decisions that backfire. For instance, you decide you want fresh, local vegetables from a farmer's market located outside of town, but you need to drive a round-trip distance of 7km to get the food. Researchers at the University of Exeter determined that this little drive was likely to cancel out all of the carbon savings that you generated from buying local.[23]

Our intuition, driven by images of dark exhaust fumes coming from fuel pipes, tells us that transporting food from around the world must be devastating to the environment. Science tells us that it's bad, but not the biggest thing to worry about in terms of food and the climate – focus on changing what you eat as your top priority. That's the best way to reduce greenhouse gases from agriculture, and the best way to ensure we have enough food for every person on the planet.

3

Collective Action

Throughout this book we've looked at actions that could take chunks out of your carbon footprint. But if you're only subtracting from your own footprint, there's always going to be a limit to how effective you can be. In this chapter we're looking at actions that can go beyond your own emissions and spur change in your friends, your neighbours and your society. Let's start with the big picture and examine some of the basics of politics and the environment.

- The more environmental protests there are in a US state and the more people donate to environmental groups the lower greenhouse gas emissions tend to be.[1]
- The stronger the "Green Parties" are in a country, the lower the airborne pollution is likely to be.[2,3]

- Individual power plants have lower emissions if there are active environmental groups nearby,[4] because these groups do things like protest until a utility company retrofits the power plant to make it less polluting.
- When there is no local opposition to infrastructure and energy projects they tend to receive government approval regardless of whether the project poses serious environmental risks.[5]

I could add more to this list, but the point is that being politically active works to protect the environment, at least in democratic countries.[3]

So what are the top actions you can take to get involved and make a difference for the climate?

Unfortunately, it's not as easy to measure the effectiveness of an action like voting in the same way we can measure other actions in this book, like going vegan. But if we want to know how to persuade politicians to make climate-friendly policy, one thing we could at least do is ask the politicians. A study from 2012 did just that – Members of Parliament in eight European countries were asked to rank different forms of political participation by impact on policy and lawmaking.[6] You can see the ten actions, ranked from highest to lowest, below, with my interpretation of how they might apply in the climate context shown in brackets:

1 Vote.

2 Get active in a political party
(or a climate organization).

3 Create media attention.

4 Write a letter to a politician.

5 Write an email to a politician.

6 Demonstrate (go to a climate protest).

7 Sign a petition.

8 Have internet discussions.

9 Boycott (and divest).

10 Perform illegal acts.

Now these aren't conclusive, but they do give us a great starting point. We're going to look at a few of the top options in a bit more detail, and since I'm a researcher and not a criminal mastermind, I won't be giving suggestions for number 10.

Vote

Every once in a while there's an opportunity
to vote directly for actual policies that
can have an effect on climate change. In
the United States there have been ballot
initiatives for Renewable Portfolio Standards,
which require utility companies to obtain
a certain amount of their energy from
renewable sources. In 2018, Washington
State also had a ballot initiative attempting to
put a price on carbon, while in Europe there
is the occasional vote on the use of nuclear
energy. But most of the time, voting for
climate action means voting for candidates
who you believe will take strong action on
climate change. While it is great to persuade
your elected officials to take action once
they're already in power, it is much, much
easier to elect a representative who doesn't

need nudging. So vote for candidates who take the climate seriously.

Get active in an organization

Joining a political party gives you influence over party platforms and leadership decisions that normal voters don't have – which is probably why MPs rated this action so highly. But when we're talking about climate impact instead of just political impact, it makes just as much sense to think about joining a climate organization.

When you're on your own it can be hard to keep track of all the ways to speak out for the climate. But an organization can let you know when and where events are happening and how to join in. They also amplify your voice by synchronizing messages with like-minded individuals. An elected official is probably going to be more responsive to a dozen phone calls in a day from an organized group than a lone voice. Finally, European MPs rated media attention as a fairly important part of influencing their decisions, and it's way easier to achieve media attention as part of a group with a strategy. Some groups around the world that are pushing for change include:

- 350.org (international)
- Climate Action Network (international)
- Citizens' Climate Lobby (international)
- Extinction Rebellion (international)
- Hope for the Future (UK)
- Sunrise Movement (US)
- RepublicEN (US)
- Quit Coal (Australia).

There is something satisfying and very human about connecting with others and working towards a common goal. Climate groups have different goals and different tactics. Some are trying to stop pipelines by holding protests and rallies, others are demanding political action by going on strike. So check out their websites, see what groups are aligned with your values and meet close to you and try one out.

Create media attention/ demonstrate

Although it's easier to get media attention and attend a demonstration when someone else is organizing it, you too can be the person leading the way. Greta Thunberg was just one of thousands of school students in Sweden, but her choice to go on strike from school and to protest for a better future started a wave of action that has resulted in thousands of students striking in hundreds of cities. In the UK, the Extinction Rebellion movement drew widespread media attention by shutting down major bridges in London, resulting in the British Parliament declaring a climate emergency.

Broader awareness can also be raised not just by protesting or going on strike but through more creative endeavours like public art installations.

Write to a politician

Whether writing a letter on your own or as part of an organized group effort, contacting an elected official is a great way to remind them that their voters care about climate change. You don't have to be an expert on policy – that's their job – you just need to tell them that you're concerned about climate change and you want to see strong action on their part. And you don't have to be someone who voted for them. If you did and you want to mention this, that's fine, but it's not necessary. If you are a constituent, they will care about your opinion.

How should you reach out? Phone calls and in-person meetings are great because they're personal, whereas emails and letters are good because they leave a traceable record. If you'd prefer to write a letter or send an email, remember these two important points:

Always start by letting them know that you are one of their constituents (politicians are much more responsive to their own voters).

End by saying that you're looking forward to hearing back from them (they should take the time to answer your questions and show that they've received your message).

Here's an example to use as a starting point:

Dear [Name of Elected Official here]

I am writing to you as one of your constituents because I am concerned about the serious consequences that climate change is already having on our [nation/region/ city]. Climate change has always been viewed as an environmental problem, but we are now realizing that it is also an economic and public health issue. [Insert specific example of something happening in your area, and preferably something that affects you and your neighbours.]

Because of the dramatic consequences of climate change, I am asking you to represent me by taking strong action to address this issue. [Ideally suggest something here that your elected official has control over, such as

green jobs, cycling infrastructure, pipeline or
airport runway construction, carbon pricing
or air pollution.]

I look forward to hearing back from you
on this critical issue,

Sincerely,

[Your name here]

So when should you write to your elected
official? Right now is an opportune moment.
We don't really have time to waste. You
could go a step further and send a batch of
letters with a few friends or neighbours. If
your elected official receives a similar (but
ideally not identical) message from a group
of different people at the same time they are
more likely to take action. I like to send a
message after every major election and any
time I attend a protest. The latter option can be

especially beneficial because you can reference a news article about the protest so that your representative knows that you're not alone.

If you are a little curious about what policies to support or projects to oppose when contacting a representative, here's a blunt thought on our global carbon budget from the executive director of the International Energy Agency: "We have no room to build anything that emits CO_2."[7] This basically means no new airport runways, coal power plants, natural gas power plants or pipelines if we want to stay within a safe temperature range on our planet. If that sounds dramatic to you then you are starting to comprehend just how desperate our situation really is. Contact your representative soon and make it a habit.

Tip: Don't forget that city/town councillors also have power over policies that affect the climate and may have more time to hear your concerns than a national representative!

Boycott/divest

Because so many of the decisions in our world are guided by money, it would be irresponsible to skip over this topic. In the case of climate change, an easy way to "boycott" fossil fuel companies is by not buying their products. That means living car-free, avoiding air travel, and reducing your home energy consumption. You can also be proactive by donating to an organization that fights for climate action, though there is also another way to leverage finance for the climate.

Divestment acts in the opposite way of an investment; you pull your financial support away from a firm for social reasons, using your money to send a message that encourages change.

Divestment was most famously used to pressure South Africa into ending apartheid, but the practice has been taken up by anti-fossil fuel campaigners. Bill McKibben of 350.org puts it best: "If it's wrong to wreck the climate then it's wrong to profit from that wreckage."

Without realizing it, many of us attend schools that invest in coal companies or pay into pensions that support the oil and gas industry – but we don't have to.

Fossil fuel companies hold the rights to resources that cannot be used if we are to prevent dangerous levels of planetary warming. A study in the prestigious journal *Nature* found that "a third of oil reserves, half of gas reserves and over 80 per cent of current coal reserves should remain unused from 2010 to 2050 in order to meet the target of 2°C".[8]

The divestment movement has gone global. The University of Glasgow, the World Council of Churches, the City of Oslo and over 1000 other institutions have made various commitments to divest. If you attend a university there's a good chance you can find a movement on your campus encouraging your institution to divest its endowment from fossil fuels. If you're a graduate (especially one who donates) you can also apply pressure in this regard. If you contribute to a pension fund, then you can join the movement by petitioning your pension fund manager to divest. And if you invest your own money you can have a close look at your portfolio and make sure that your money is in line with your morals.

Talk about it!

One of the many obstacles blocking progress
on climate change is the fact that people
don't want to talk about it. This ranges from
members of the public all the way to national
politicians.[9] Even people who believe in
climate change may be hesitant to speak
out, thinking that they are in the minority
even when they are not. Some researchers
describe this as a self-reinforcing "spiral
of silence".[10] It's no wonder that one of the
most well-respected climate researchers
in the world, Katharine Hayhoe, says that
the first thing you can do to fight climate
change is to talk about it, focusing on values
you have in common, whether those be
parenting, outdoor activities or your local
community, and connecting these things
with climate change.[11]

I grew up in rural Ontario, and even now climate change can be a frosty subject to bring up. When I tell people that's what I study, I sometimes wince inside wondering what the reaction is going to be. But I was more than a little inspired recently by an encounter on a city bus. A construction worker came through and sat behind me. Fluorescent safety vest, real confident, he popped a couple windows open on his way to the back and sat down with his arms spread on the seats behind him. He struck up a conversation with the person next to him; I think they were talking about the weather. The person next to him said, "the thing about climate change…", and the construction worker interrupted emphatically to say, "the thing about climate change is that it is REAL" and he went on to share his opinions on the seriousness of the subject. I had very much misjudged where that conversation was going.

Now that was just one anecdote, but it exemplifies a trend that has been documented by researchers. Even in America, where climate change is probably more politically polarized than anywhere else on Earth, almost 75 per cent of the population believes the Earth is warming. And yet, those like-minded people estimate that only 57 per cent of their fellow Americans think the same way that they do.[12] Your default assumption should be that people care about the climate, because that's what the numbers say. Talking about the climate helps to end the self-imposed silence.

When asked, Americans reported that the group that was most capable of convincing them to take action on global warming was their own friends and family.[16] For instance, while 27 per cent said that their significant other could persuade them to take action, only 13 per cent said the same was true for their environmental leaders and only 6 per cent for politicians. Conversation is powerful.

Have an internet discussion

Because online social media users tend to be sorted (and isolated) according to their political beliefs, attempts at breaking through these echo chambers can be difficult. For instance, one study looked at 54 million Facebook users over a five-year time frame.[13] The users were split into those who interact with conspiracy-type pages, and those who interact with science pages. When users "debunked" scientific misinformation, very few of the conspiracy-type users interacted with those posts, and when they did it seemed to spur them to interact even more with conspiracy pages afterwards. So online arguments can backfire by causing people to become even more entrenched in their own views. Now this is a new research area, and things change fast in the world of social media, so we shouldn't assume these results apply to every situation. But it does look like at least some online interactions might have the opposite of the intended effect.

Some information can penetrate a partisan haze. For instance, studies have found that 97 per cent of climate scientists agree that climate change is happening and is caused by humans.[14] And online messages that relate this fact result in increased public support for climate action, even among the sceptical.[15] That being said, if you're interested in persuading others to take climate change seriously, it might be best to stick with face-to-face conversations with people you know. And if you're labouring over a witty, well-cited reply to someone on social media you don't know, you are better off taking the time to send a message to an elected representative whose job security rests on interpreting the preferences of their constituents.

Sign petitions

The signing of petitions is rated as less effective by European politicians. Here I would differentiate between internet petitions and petitions that are formally submitted through a democratic process. Some nations have structural measures in place where petitions force acknowledgements from the government, and thereby ensure some level of productivity.

In the UK, a petition with over 100,000 signatures is considered for a debate in parliament. In Canada, any paper petition with 25 signatures can be presented to parliament by a willing MP. In these cases, all of the politicians in attendance are forced to acknowledge that there are citizens who care about the issue. Sometimes large petitions also generate media attention, which sends a signal to politicians that this is something they should pay attention to. In Switzerland, the results of a petition are even more concrete – when 100,000 citizens sign a federal popular initiative it can trigger a nationwide referendum!

So signing onto a petition can be productive, and is certainly low effort. The only thing I would caution against is signing a petition and then thinking that you've done enough.

4

Everyday Living

There are a WHOLE lot of behaviours to consider when thinking about your carbon footprint around your home.

At Home

Energy consumption

It's harder to rank actions at home and work by calculating one tCO_2e number that is accurate for all cities and countries. If you live in a country that relies on coal power, like Australia or China, choosing to reduce your energy consumption will make a huge reduction to your carbon footprint.

Energy reduction makes a much smaller impact if you're in a place where the grid relies on hydro (water) power, like Quebec, or nuclear power, like France. In the table below I've shown how many kilograms of CO_2 would come from 100 hours of watching a big-screen TV in a few nations to illustrate the differences (remember earlier that we identified 1kg of CO_2 as a 7km car drive).

Country	KgCO$_2$e per 100 hours of TV
Australia	13.5
United States	6.6
United Kingdom	4.2
Canada	2.3
New Zealand	1.8

Note that emissions will vary by region even inside a country, and that these are based on national annual averages

*Assumes a 150 W High Definition television

The large differences between energy grids might suggest to you that focusing on changing the system is more important than focusing on changing the small behaviours that only affect you. It's pretty hard for someone living in an area with lots of coal power to get their carbon footprint to a sustainable level because the electricity they use in their home is so carbon intensive. It's hard for individuals to be sustainable on our own. We need governments to make change, and the way we can do that is through collective action (page 69).

Now there are still some big-picture changes that can be made around the home, but they might not come up often. Our guiding practice in this book is to start with the most important actions and work our way down. In terms of housing, the big things you can do, in approximate order, are:

- live close to work.
- own a smaller home/share your living space.
- get renewable energy.
- weatherproof your home and purchase energy-efficient appliances.

Living close to work

Living close to work is mostly about shrinking the footprint of your commute. If you live in the area of the city or town where you also work, it is more likely that you live in an area where everything you need is closer to you so you can walk or take public transport instead of drive. Real estate is more expensive in denser areas, so if you live closer to where you work you probably have a smaller living space too. Apartment buildings, condos and rowhousesshare walls, making them easier to heat and making your personal energy consumption lower than someone living in a detached house. When you have less space, you also have fewer things, which come with their own carbon footprint as well. For these reasons, studies show that households in denser, urban environments have lower carbon footprints than households in the suburbs. But there is one issue: households in dense, urban areas sometimes have higher emissions than expected because their occupants fly more.[1] A couple of trips by plane can undo a lot of good accomplished by an otherwise sustainable lifestyle.

Shared living space

Sharing your living space is a similar way to reduce your climate impact without needing to constantly remind yourself to do something. Each person added to a household reduces *per capita* emissions by somewhere between 6 per cent[2] and 9 per cent[3] (about 1.4 tCO_2e in the United States). Larger households live with less environmental harm simply by sharing – and not just living space. They share heating, transportation, refrigerators, washing machines, cutlery and wifi, and it all adds up. I recognize that this isn't a decision that you can make every day. But it's good to know that families taking on students in a spare room and single people splitting rent with roommates are doing more than just saving money.

Renewable energy

There are a couple of ways to get renewable energy for your home. If you own your house, install solar panels on your roof. There is an up-front cost but that gets paid off over time, eventually earning you back savings and in some cases, increasing the value of your home.[4] Different jurisdictions have different incentives that can make this pay-back time longer or shorter. Research what those are as well as who to hire in your area. The benefit to the climate can be substantial. And don't forget that an action like installing solar panels is one of those contagious actions where your decision can have a ripple effect in your community. Depending on where you live there may be an option to install geothermal energy as well!

If you don't own your own house, but you still pay for your own heating and energy, you can check with your utility company and see if you can switch to renewable energy. I currently pay a small extra fee so that my share of the natural gas that heats my home is sourced from decomposing organic matter at a landfill, instead of from a fossil fuel deposit taken from the ground. Some utilities also allow you to purchase electricity derived from renewable sources (solar, wind, etc.).

On top of these changes, ideally, you'd want your money to be funding a growing renewables market that creates additional green energy that goes into the grid. Findings have shown that some energy companies are using customer money to pay for "green electricity" that already exists[5] (like an old hydroelectric dam rather than the construction of a new wind turbine). If you switch to a renewable energy source, do some research yourself to see if your money is being spent transparently on new projects that help reduce reliance on fossil fuels.

Weatherproofing and efficient appliances

If you own your own home, then you can reduce your home energy use by better weatherproofing. Here are some of the ways you can keep heat in the house during colder months:

- cavity wall insulation
- loft insulation
- double- or triple-glazed windows
- draught stoppers under doors.

Energy-efficient appliances can reduce your home electricity use. Look for energy labels when you purchase new appliances and compare with the other products available. There are a few up-and-coming technologies to consider adopting early, for example: an induction stovetop. These are more responsive and safer than gas stove tops but use electricity, so their effect on the climate is smaller (especially as nations decarbonize their energy grids). Heat pumps are more efficient than air conditioners and heaters

and can do the job of both. They move heat
indoors in the winter and outdoors in the
summer (using similar tech to a refrigerator
cooling system). Both of these technologies
are developing, so expect their costs to drop as
time goes on.

Laundry

I often get asked to recommend the most
effective actions for the climate that are also
the easiest. I really don't like the framing of
that question. It says that the future of the
planet is less important than convenience. Or
maybe it shows a failure to comprehend the
gravity of the situation we're in. Regardless,
I do recognize that people can get fatigued by
taking on too many changes. For that reason,
there are other actions that are pretty easy
to do and still conserve energy enough to

meaningfully affect your carbon footprint.
A couple of these involve laundry.

Anything that involves heating up a lot of
water will require a lot of energy. If you can,
wash your clothes in cold water as often as
possible to cut back on energy. All you have
to do is set your washing machine to cold,
and modern detergents work equally well
in cold water. Then once you've done your
laundry, air dry it on a clothes rack or outside
on a washing line rather than using a dryer.
This takes a little more work (and if you have
moths or other pests then you may need to
wash clothes with hot water or use a dryer to
keep them at bay), but it's less hard on your
clothes, so they will last longer. Washing with
cold water and air drying clothes combined
save a little more than 0.4 tCO_2e each year,
assuming you do them all year long.[6]

Lighting

We often think of turning off the lights as being an environmentally aware action. But these more visible behaviours (e.g. lighting) are much less effective than one-time behaviours that are hidden out of sight (e.g. installing a more efficient water heater).[7] If you want to reduce the energy you use for lighting, switch to efficient lightbulbs, like LEDs.

Heating

Setting your thermostat to be a couple degrees cooler in the winter, and your air conditioner to be a couple degrees warmer in the summer, is a worthwhile change. If you can substitute a fan for the AC that's an even better solution for summer heat.

Energy saving – why it matters wherever you live

At the start of this chapter I mentioned that it's hard to quantify behavioural changes involving household electricity because different energy grids use different amounts of solar, nuclear, coal and so forth. But if you're in an area with really clean energy, that doesn't mean saving energy is pointless.

Provinces, countries and utility companies must make choices about supply and demand. If people live in bigger homes, buy bigger television screens or switch to electric vehicles then utility companies need to provide more electricity. This might mean building a new dam and flooding people and animals out of their homes. It could also mean leaving a coal power plant open longer if renewable energy isn't improving quickly enough to match the rise in demand.

This is one of the reasons why utility companies often run energy-saving programmes and give feedback to their customers about their energy use. They don't want to take on a risky new infrastructure project; they'd rather help their customers save energy. If they believe that every household's energy use adds up, then you should too.

At Work

Nobody can tell you exactly what the best thing to do for climate change is at your workplace. But you are likely to be more effective if you can promote structural changes or create new company policies that have far-reaching effects. It is one thing if you can find a way to avoid driving to work. But it is quite another if you can change company policy so that, for example, everyone eligible for free parking has the option to cash out the value of that monthly parking space. One requires you to be diligent about your travel, while the other incentivizes an entire company to carpool, bike or take public transport.

Company changes that could have a large impact:

- If your company owns a vehicle fleet, find a way to make them more efficient (route optimization, purchasing/leasing more fuel-efficient vehicles, reducing the number of trips taken).
- Avoid air travel for staff by getting good videoconferencing software/equipment, and rewarding employees with perks other than trips abroad.
- Install electric vehicle charging stations in the company parking lot and provide cycle storage with showers and changing rooms to encourage cycling.
- When company equipment is upgraded, choose the most energy-efficient technology to replace it; for example, laptops are more

energy efficient than PCs, and can also help cut paper use.

- Create a policy with the procurement team (if you have one). If not, ensure that all those who purchase items for the company embed environmental (and social) considerations into the purchase – to make sure that the business is getting the best deal in the long term.

Really though this is about taking a long moment to stop and think about what sort of decisions you can make at work.

If you're working an entry-level job in a restaurant or an office, you may not have a lot of say over larger structural changes. In that case, tips to be more sustainable look a lot like they might at home:

- Set the AC so that nobody needs to wear a jumper in July (and if management insists on wearing suit jackets then make a case about why energy conservation is important).
- Set your photocopier to power-save mode and secure print.
- If you're the last one out of the office, switch off the lights, computer monitors, printers, etc., or ask whoever stays latest (e.g. security staff) to do it.
- Try to reduce the amount of paper needed in meetings – does everyone need their own agenda and meeting notes, or can you share/have a digital version on a screen?

Research also suggests that pro-environmental behaviours in the workplace are contagious. Advocating for green values and choices establishes norms that change workplace behaviour, and are especially powerful when coming from leaders.[8] But if you're not a leader you can at least try to take on that role. Consider forming a "green team", or sustainability committee, with like-minded individuals in the company. The City of Portland has put together some great resources and tips on what this could look like.[9] A green team could tackle anything from an annual week-long cycling commute competition to a programme aiming to reduce on-site paper consumption. If you have big plans and need to sell your vision to management, here are a couple of ideas:

- Remind management that as the public pushes the government to address environmental concerns, regulations

will tighten. Companies can find success
by establishing practices that anticipate
changes in legislation, making them leaders
in the field.[10]

- Whether in the form of car fuel or
  electricity, energy costs money. This means
  that any conservation efforts are likely to
  save a business money as well.
- Green is a good look. Both shareholders
  and members of the public care about
  corporate sustainability, so any efforts
  that a company takes should be proudly
  communicated to both of these groups!

Eeny, Meeny, Miny, Moe

We're faced with dozens of decisions each day. These choices aren't going to change the world, and some of them won't have that big of an effect on your carbon footprint. But if you care about climate change, and you don't like waste, then there are some questions you might want to have answered:

- *Blow dry my hands or use a paper towel?* The best answer is to wipe your hands on your trousers and move on! Often the best answers for the environment aren't fancy technology, they're plain and simple.

- *Paper or plastic?* For grocery bags it all depends on how many times you reuse them. Plastic bags have a smaller carbon footprint than paper or even canvas bags, but you can't get as many uses out of them. The key here is to get one canvas bag and reuse it a lot – aim for over 100 times.[11]

- *Cat or dog?* I'm a dog person myself, but in terms of carbon footprint cats have an edge. Most of the footprint of an animal comes from what they eat, and since dogs are usually bigger, they usually eat more. But cat food tends to have more protein,[12] so on a gram for gram level it is likely to be more carbon intensive. Outdoor and free-range cats are prolific bird killers, estimated at 1.3–4 billion birds every year in the United States,[13] which is not good! Other pets (rabbits, turtles) have a smaller footprint because of their size and diet (no meat). You can also reduce the footprint of your pet by not overfeeding them.[14]

- *Ecolabel cleaning product or bleach?*
  I'm going to pull a fast one again and say
  neither. Have you tried white vinegar? Or
  vinegar paired with baking soda? These
  are options that don't require the same
  industrial intensity as something like bleach.
- *E-reader or hardcover books?* Depending
  on the study, the average book has a carbon
  footprint of around 2 $kgCO_2e$ per kilogram
  of weight.[15] The physical copy you have in
  your hands, for instance, weighs much less
  than a kilogram – so if these pages inspire
  you to forego eating a single hamburger
  you will have likely come out ahead.
  An e-reader has a much larger carbon
  footprint (20–40$kgCO_2e$ per e-reader), so
  to offset the e-reader you would need to *not*
  purchase about 40 physical books.[15] The
  best option is to borrow from the library.

Bigger, Newer, More Is Not Better

Hidden among the noise of household changes are all of the hundreds of decisions that contribute to an over-consumptive lifestyle. Buying a new mobile phone when your old one works perfectly well. Having a closet full of clothes, many of which you don't wear. Living in a five-bedroom home occupied by just yourself and your partner. I can't put a number to the sum of all these actions, but they can add up. As a society we need to move away from retail therapy and embrace lives that find meaning outside of consumption.

Conclusion:
How It All
Adds Up

You might have seen recent press around individuals choosing to not have children for environmental reasons. The decision of how many children to have is deeply personal, and each person is going to weigh many different considerations when they make this choice (and climate change might be one thing that they want to consider). So let's look at what we know.

The more people there are on the planet the more resources we will consume and the harder it is to reach climate targets. One study looked at the carbon legacy of having an additional child, finding that under current conditions, each child in the United States can be expected to add 9441 tCO_2e, both through their own actions and the actions of the children they are likely to have themselves.[1]

But that number decreases by up to 17 times in a scenario where society acts quickly to drastically reduce emissions. We want to be in a society where having more children does not mean such a huge burden on the environment. And you can help to make it happen.

In terms of reaching our short-term climate goals (which is the purpose of this book), population is not going to be the difference maker. We already have enough people on the planet to blow past the climate targets that the international community has agreed upon, and so what is really needed in the here and now are changes in technology and consumption. You can affect consumption by choosing how you live (what you eat, how you get around) and you can affect both technology and consumption by putting pressure on your government through collective action.

Final thoughts

As you wrap up this book, I want to leave you with a few parting thoughts concerning what scientists understand about change – both how individuals change and how society changes.

The first thought is something of a warning. There is a phenomenon in social science known as "moral licensing". It's the idea that doing one thing that's good might give people the feeling that they can then go on to do something that's bad: well I went for a run earlier, so it's fine to eat a slice of cake. This does not work so well when dieting and does not work for climate change either.

There is also some evidence of moral licensing happening with the environment. In one study, people who received feedback on their water usage, and went on to lower it, were found to increase their home energy use.[2] This is not ideal.

Moral licensing is somewhat related to the "rebound effect". In the rebound effect, energy efficiency improvements lead to greater energy use overall. Take an example: you buy a hybrid vehicle that uses less petrol. But when you have a more fuel-efficient vehicle, you can drive it further for the same cost of petrol, so you drive more. Now the environmental savings are gone. Or let's say you switch to a plant-based diet and begin saving money because meat is more expensive than beans. You might not even realize why you have the extra money, but at the end of the year you go and spend your savings on a flight to Rome. Then the good that you did for the environment has been largely erased.

We can avoid this by focusing on those high-impact actions: **live car-free, avoid air travel *and* eat a plant-based diet** and you're less likely to get that kind of backfire effect.

In the same way that researchers study the interesting and unpredictable impact that an individual's changes can have, researchers also investigate group dynamics. Tipping points are an example of an interesting social dynamic with big implications for the climate movement.

Researchers believe that tipping points are more likely where a silent majority agrees with the goals of the social movement. This is the case for climate change generally, as well as for movements like the Fridays for Future protests.[3] If you are at a school where only a dozen students go on strike for the climate every Friday, it may be socially "costly" to join in because it's not a very popular activity. But as more and more students take the day to protest for the sake of the future, the social cost of joining becomes lower. At some point, the movement reaches a tipping point where it may be more difficult *not* to join, and suddenly you see thousands of students in the streets, media attention spikes, and politicians are forced to reckon with the issue of climate change.

You can imagine the same pattern playing out in other areas. At first only a small portion of a community eats a plant-based diet. But some people join for health reasons while others see a film about the industrial meat industry and decide to try a new diet as a result. That little nudge might create enough of a market for the first vegetarian restaurant in town to open and soon every fast food joint

needs to at least offer a veggie burger. Now
the costs of acting are much lower because
you can go out with your friends for a meal
and always find a vegetarian option, but
also because what is socially acceptable has
changed. Our political and social systems may
reach tipping points sooner than we realize,
and that's a good thing, because there's no
time left to waste.

What Actions Will You Take?

Here's a round-up of all the ideas in this book that will make an impact on your carbon emissions. How many will you commit to do?

☐ Take one fewer flight a year

☐ Take a train instead of a plane

☐ Take a staycation

☐ Live car free (2.4tCO_2e per year)

☐ Switch your diesel or petrol car to electric (1.25tCO_2e per year)

☐ Use public transport

☐ Go vegan (0.9tCO_2e)

☐ Go vegetarian (0.8 tCO_2e)

☐ Eat pescatarian

☐ Make your vote count

☐ Join an organization

☐ Write to a politician

☐ Demonstrate (go to a protest)

☐ Sign a petition

☐ Start a conversation with friends
and family

☐ Divest

☐ Live close to work

☐ Own a smaller home/share your living
space (about 1.4 tCO_2e)

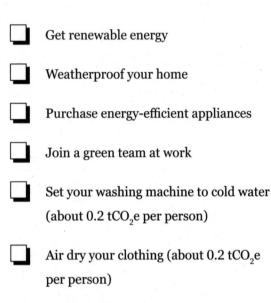

☐ Get renewable energy

☐ Weatherproof your home

☐ Purchase energy-efficient appliances

☐ Join a green team at work

☐ Set your washing machine to cold water (about 0.2 tCO_2e per person)

☐ Air dry your clothing (about 0.2 tCO_2e per person)

Notes

Introduction

[1] Caldeira, K. "How much ice is melted by each carbon dioxide emission", <https://kencaldeira.wordpress.com/2018/03/24/how-much-ice-is-melted-by-each-carbon-dioxide-emission/> (2018).

[2] Graziano, M. and Gillingham, K. "Spatial patterns of solar photovoltaic system adoption: The influence of neighbors and the built environment." *Journal of Economic Geography* 15 (2014): 815–839, doi:10.1093/jeg/lbu036.

[3] Sparkman, G. and Walton, G. M. "Dynamic norms promote sustainable behavior, even if it is counternormative." *Psychological Science* 28 (2017): 1663–1674, doi:10.1177/0956797617719950.

[4] Schultz, P. W., Nolan, J. M., Cialdini, R. B., Goldstein, N. J. and Griskevicius, V. "The constructive, destructive, and reconstructive power of social norms." *Psychological Science* 18 (2007): 429–434, doi:10.1111/j.1467-9280.2007.01917.x.

5 Steinfeld, H. *et al. Livestock's long shadow: environmental issues and options.* (Food & Agriculture Org., 2006).

6 Lee, D. S. *et al.* "Aviation and global climate change in the 21st century." *Atmospheric Environment* 43 (2009): 3520–3537, doi:10.1016/j.atmosenv.2009.04.024.

7 Lacroix, K. "Comparing the relative mitigation potential of individual pro-environmental behaviors." *Journal of Cleaner Production* 195 (2018): 1398–1407, doi:10.1016/j.jclepro.2018.05.068.

8 European Commission. "Greenhouse gas emissions per capita", <https://ec.europa.eu/eurostat/tgm/table.do?tab=table&init=1&language=en&pcode=t2020_rd300&plugin=1> (2018).

9 Data from https://di.unfccc.int/detailed_data_by_party, with total GHG emissions divided by population size.

10 Girod, B., Van Vuuren, D. P. and Hertwich, E. G. "Global climate targets and future consumption level: an evaluation of the required GHG intensity." *Environmental Research Letters* 8 (2013): 014016, doi:10.1088/1748-9326/8/1/014016.

Getting Around

1 Data adapted from: Le Quéré, Corinne, et al. "Towards a culture of low-carbon research for the 21st Century."

Tyndall Centre for Climate Change Research, Working Paper 161 (2015).

[2] Eijgelaar, E., Thaper, C. and Peeters, P. "Antarctic cruise tourism: the paradoxes of ambassadorship, 'last chance tourism' and greenhouse gas emissions." *Journal of Sustainable Tourism* 18 (2010): 337–354, doi:10.1080/09669581003653534.

[3] Cames, M., Graichen, J., Siemons, A. and Cook, V. *Emission reduction targets for international aviation and shipping.* Policy Department A: Economic and Scientific Policy, European Parliament, B-1047 Brussels (2015).

[4] Schäfer, A. W. *et al.* "Technological, economic and environmental prospects of all-electric aircraft." *Nature Energy* 4 (2018): 160–166, doi:10.1038/s41560-018-0294-x.

[5] International Civil Aviation Organization. *Conference on aviation and alternative fuels: Aviation biofuels efficiency in terms of CO_2 emissions reductions.* (2017).

[6] Tollefson, J. "Humans are driving one million species to extinction." *Nature* 569 (2019): 171, doi:10.1038/d41586-019-01448-4.

[7] Bows-Larkin, A. "All adrift: aviation, shipping, and climate change policy." *Climate Policy* 15 (2015): 681–702, doi:10.1080/14693062.2014.965125.

8   Westlake, S. *A Counter-Narrative to Carbon Supremacy: Do Leaders Who Give Up Flying Because of Climate Change Influence the Attitudes and Behaviour of Others?* Available at SSRN 3283157 (2017), doi:10.2139/ssrn.3283157.

9   Frank, L. D., Andresen, M. A. and Schmid, T. L. "Obesity relationships with community design, physical activity, and time spent in cars." *American Journal of Preventive Medicine* 27 (2004): 87–96, doi:10.1016/j.amepre.2004.04.011.

10 Greenstone, M. and Fan, C. Q. *Introducing the Air Quality Life Index.* (Energy Policy Institute, University of Chicago, 2018).

11 Savage, I. "Comparing the fatality risks in United States transportation across modes and over time." *Research in Transportation Economics* 43 (2013): 9–22, doi:10.1016/j.retrec.2012.12.011.

12 Lancée, S., Veenhoven, R. and Burger, M. "Mood during commute in the Netherlands: What way of travel feels best for what kind of people?" *Transportation Research Part A: Policy and Practice* 104 (2017): 195–208, doi:10.1016/j.tra.2017.04.025.

13 Transport, D. f. *The Road to Zero: Next steps towards cleaner road transport and delivering our Industrial Strategy*, <https://assets.publishing.service.gov.uk/government/uploads/system/uploads/

attachment_data/file/739460/road-to-zero.pdf>
(2018).

[14] Premier, O. o. t. "Provincial government puts
B.C. on path to 100% zero-emission vehicle
sales by 2040", <https://news.gov.bc.ca/
releases/2018PREM0082-002226> (2018).

[15] Knudsen, C. and Doyle, A. "Norway powers ahead
(electrically): over half new car sales now electric
or hybrid", <https://www.reuters.com/article/us-
environment-norway-autos/norway-powers-ahead-
over-half-new-car-sales-now-electric-or-hybrid-
idUSKBN1ES0WC> (2018).

[16] Barrington-Leigh, C. and Millard-Ball, A. "More
connected urban roads reduce US GHG emissions."
Environmental Research Letters 12 (2017): 044008,
doi:10.1088/1748-9326/aa59ba.

What We Eat

[1] Poore, J. and Nemecek, T. "Reducing food's
environmental impacts through producers and
consumers." *Science* 360 (2018): 987–992, doi:
10.1126/science.aaq0216.

[2] Springmann, M. *et al.* "Options for keeping the food
system within environmental limits." *Nature* 562
(2018): 519–525, doi:10.1038/s41586-018-0594-0.

3   Tollefson, J. "Humans are driving one million species to extinction." *Nature* 569 (2019): 171, doi:10.1038/d41586-019-01448-4.

4   Veeramani, A., Dias, G. M. and Kirkpatrick, S. I. "Carbon footprint of dietary patterns in Ontario, Canada: A case study based on actual food consumption." *Journal of Cleaner Production* 162 (2017): 1398–1406, doi: 10.1016/j.jclepro.2017.06.025.

5   Hoolohan, C., Berners-Lee, M., McKinstry-West, J. and Hewitt, C. "Mitigating the greenhouse gas emissions embodied in food through realistic consumer choices." *Energy Policy* 63 (2013): 1065–1074, doi:10.1016/j. enpol.2013.09.046.

6   United Nations. *World Population Prospects: The 2017 Revision, Key Findings and Advance Tables.* (Department of Economic and Social Affairs, Population Division, 2017).

7   Cassidy, E. S., West, P. C., Gerber, J. S. and Foley, J. A. "Redefining agricultural yields: from tonnes to people nourished per hectare." *Environmental Research Letters* 8 (2013): 034015, doi:10.1088/1748-9326/8/3/034015.

8   Shepon, A., Eshel, G., Noor, E. and Milo, R. "The opportunity cost of animal based diets exceeds all food losses." *Proceedings of the National Academy of Sciences* 115 (2018): 3804–3809, doi:10.1073/pnas.1713820115.

9 Röös, E., Patel, M., Spångberg, J., Carlsson, G. and Rydhmer, L. "Limiting livestock production to pasture and by-products in a search for sustainable diets." *Food Policy* 58 (2016): 1–13, doi:10.1016/j. foodpol.2015.10.008.

10 Myhre, G. et al. *Anthropogenic and Natural Radiative Forcing. 731* (IPCC, Cambridge, United Kingdom, 2013).

11 Bryngelsson, D., Wirsenius, S., Hedenus, F. and Sonesson, U. "How can the EU climate targets be met? A combined analysis of technological and demand-side changes in food and agriculture." *Food Policy* 59 (2016): 152–164, doi:10.1016/j.foodpol.2015.12.012 .

12 Tilman, D. and Clark, M. "Global diets link environmental sustainability and human health." *Nature* 515 (2014): 518–522, doi:10.1038/ nature13959.

13 Garnett, T. "Livestock-related greenhouse gas emissions: impacts and options for policy makers." *Environmental Science & Policy* 12 (2009): 491–503, doi:10.1016/j.envsci.2009.01.006.

14 Harwatt, H., Sabaté, J., Eshel, G., Soret, S. and Ripple, W. "Substituting beans for beef as a contribution toward US climate change targets." *Climatic Change* 143 (2017): 261–270, doi:10.1007/s10584-017-1969-1.

15 Walker, P., Rhubart-Berg, P., McKenzie, S., Kelling, K. and Lawrence, R. S. "Public health implications of meat production and consumption." *Public Health Nutrition* 8 (2005): 348–356, doi:10.1079/PHN2005727.

16 Melina, V., Craig, W. and Levin, S. "Position of the Academy of Nutrition and Dietetics: vegetarian diets." *Journal of the Academy of Nutrition and Dietetics* 116 (2016): 1970–1980, doi:10.1016/j.jand.2016.09.025.

17 United States Department of Agriculture. *USDA Coexistence Fact Sheets Soybeans.* (Office of Communications, Washington, DC, 2015).

18 Spencer, E. A., Appleby, P. N., Davey, G. K. and Key, T. J. "Diet and body mass index in 38 000 EPIC-Oxford meat-eaters, fish-eaters, vegetarians and vegans." *International Journal of Obesity* 27 (2003): 728–734, doi:10.1038/sj.ijo.0802300.

19 Beezhold, B. L. and Johnston, C. S. "Restriction of meat, fish, and poultry in omnivores improves mood: A pilot randomized controlled trial." *Nutrition Journal* 11:9 (2012), doi:10.1186/1475-2891-11-9.

20 Messina, V. "Nutritional and health benefits of dried beans." *The American Journal of Clinical Nutrition* 100 (2014): 437S–442S, doi:10.3945/ajcn.113.071472.

21 Quested, T. E., Marsh, E., Stunell, D. and Parry, A. D. "Spaghetti soup: The complex world of food waste

behaviours." *Resources, Conservation and Recycling* 79 (2013): 43-51, doi:10.1016/j.resconrec.2013.04.011.

[22] Weber, C. L. and Matthews, H. S. "Food-miles and the relative climate impacts of food choices in the United States." *Environmental Science & Technology* 42 (2008): 3508–3513, doi:10.1021/es702969f.

[23] Coley, D., Howard, M. and Winter, M. "Local food, food miles and carbon emissions: A comparison of farm shop and mass distribution approaches." *Food Policy* 34 (2009): 150–155, doi:10.1016/j.foodpol.2008.11.001.

Collective Action

[1] Muñoz, J., Olzak, S. and Soule, S. "Going Green: Environmental Protest, Policy and CO2 Emissions in US States, 1990-2007." *Sociological Forum* 33 (2018): 403-421, doi: 10.1111/socf.12422

[2] Neumayer, E. "Are left-wing party strength and corporatism good for the environment? Evidence from panel analysis of air pollution in OECD countries." *Ecological Economics* 45 (2003): 203–220, doi:10.1016/S0921-8009(03)00012-0.

[3] Bernauer, T. and Koubi, V. "Effects of political institutions on air quality." *Ecological Economics* 68 (2009): 1355–1365, doi:10.1016/j.ecolecon.2008.09.003.

4   Grant, D. and Vasi, I. "Civil Society in an Age of
    Environmental Accountability: How Local Environmental
    Nongovernmental Organizations Reduce US Power
    Plants' Carbon Dioxide Emissions." *Sociological Forum*
    32 (2017): 94-115, doi: 10.1111/socf.12318

5   McAdam, D. and Boudet, H. *Putting social movements
    in their place: Explaining opposition to energy
    projects in the United States, 2000–2005.* (Cambridge
    University Press, 2012).

6   Hooghe, M. and Marien, S. "How to reach members of
    Parliament? Citizens and members of Parliament on
    the effectiveness of political participation repertoires."
    Parliamentary Affairs, 67 (2014): 536–560,
    doi:10.1093/pa/gss057.

7   Vaughan, A. "World has no capacity to absorb new fossil
    fuel plants, warns IEA", <https://www.theguardian.
    com/business/2018/nov/13/world-has-no-capacity-to-
    absorb-new-fossil-fuel-plants-warns-iea> (2018).

8   McGlade, C. and Ekins, P. "The geographical
    distribution of fossil fuels unused when limiting global
    warming to 2 [deg] C." *Nature* 517 (2015): 187–190,
    doi:10.1038/nature14016.

9   Willis, R. "Constructing a 'Representative Claim' for
    Action on Climate Change: Evidence from Interviews
    with Politicians." *Political Studies* 66 (2018): 940–958,
    doi:10.1177/0032321717753723.

[10] Geiger, N. and Swim, J. K. "Climate of silence: Pluralistic ignorance as a barrier to climate change discussion." *Journal of Environmental Psychology* 47 (2016): 79–90, doi:10.1016/j.jenvp.2016.05.002.

[11] Peters, A. "3 steps to nudge climate skeptics toward action", <https://www.fastcompany.com/90274991/3-steps-to-nudge-climate-skeptics-toward-action> (2018).

[12] Witte, M. D. "Public support for climate policy remains strong", <https://earth.stanford.edu/news/public-support-climate-policy-remains-strong#gs.51u3kg> (2018).

[13] Leiserowitz, A., Maibach, E., Roser-Renouf, C. & Feinberg, G. *How Americans communicate about global warming in April 2013*. Yale University and George Mason University. New Haven, CT: Yale Project on Climate Change Communication (2013). <https://climatecommunication.yale.edu/publications/how-americans-communicate-about-global-warming-april-2013/>

[14] Zollo, F. et al. "Debunking in a world of tribes." *PLoS One* 12 (2017): e0181821, doi:10.1371/journal.pone.0181821.

[15] Cook, J. et al. "Consensus on consensus: a synthesis of consensus estimates on human-caused global warming." *Environmental Research Letters* 11 (2016): 048002, doi:10.1088/1748-9326/11/4/048002.

16 van der Linden, S. L., Leiserowitz, A. A., Feinberg, G. D. and Maibach, E. W. "The scientific consensus on climate change as a gateway belief: Experimental evidence." *PLoS One* 10 (2015): e0118489, doi:10.1371/journal. pone.0118489.

Everyday Living

1 Heinonen, J., Jalas, M., Juntunen, J. K., Ala-Mantila, S. and Junnila, S. "Situated lifestyles: I. How lifestyles change along with the level of urbanization and what the greenhouse gas implications are—a study of Finland." *Environmental Research Letters* 8 (2013): 025003, doi:10.1088/1748-9326/8/2/025003.

2 Fremstad, A., Underwood, A. and Zahran, S. "The environmental impact of sharing: household and urban economies in CO2 emissions." *Ecological Economics* 145 (2018): 137–147, doi:10.1016/j. ecolecon.2017.08.024.

3 Underwood, A. and Fremstad, A. "Does sharing backfire? A decomposition of household and urban economies in CO2 emissions." *Energy Policy* 123 (2018): 404–413, doi:10.1016/j.enpol.2018.09.012.

4 Hoen, B., Wiser, R., Thayer, M. and Cappers, P. "Residential photovoltaic energy systems in California: the effect on home sales prices." *Contemporary Economic Policy* 31 (2013): 708–718, doi:10.1111/ j.1465-7287.2012.00340.x.

5 Hast, A., Syri, S., Jokiniemi, J., Huuskonen, M. and Cross, S. "Review of green electricity products in the United Kingdom, Germany and Finland." *Renewable and Sustainable Energy Reviews* 42 (2015): 1370–1384, doi:10.1016/j.rser.2014.10.104.

6 Wynes, S. and Nicholas, K. "The climate mitigation gap: Education and government recommendations miss the most effective individual actions." *Environmental Research Letters* 12 (2017): 074024, doi:10.1088/1748-9326/aa7541.

7 Attari, S. Z., DeKay, M. L., Davidson, C. I. and Bruine de Bruin, W. "Public perceptions of energy consumption and savings." *Proceedings of the National Academy of Sciences* 107 (2010): 16054–16059, doi:10.1073/pnas.1001509107.

8 Kim, A., Kim, Y., Han, K., Jackson, S. E. and Ployhart, R. E. "Multilevel influences on voluntary workplace green behavior: Individual differences, leader behavior, and coworker advocacy." *Journal of Management* 43 (2017): 1335–1358, doi:10.1177/0149206314547386.

9 Portland, C. o. *Sustainability at work: Green Team Guide*, <https://www.portlandoregon.gov/sustainabilityatwork/article/497862>

10 Moxen, J. and Strachan, P. *Managing green teams: environmental change in organisations and networks.* (Routledge, 2017).

[11] Edwards, C. and Fry, J. M. *Life cycle assessment of supermarket carrier bags.* Environment Agency, Horizon House, Deanery Road, Bristol, BS1 5AH (2011).

[12] Rushforth, R. and Moreau, M. *Finding Your Dog's Ecological 'Pawprint': A Hybrid EIO-LCA of Dog Food Manufacturing. Center for Earth Systems Engineering and Management* (2013).

[13] Loss, S. R., Will, T. and Marra, P. P. "The impact of free-ranging domestic cats on wildlife of the United States." *Nature Communications* 4 (2013): 1396, doi:10.1038/ncomms2380.

[14] Okin, G. S. "Environmental impacts of food consumption by dogs and cats." *PLoS One* 12 (2017): e0181301, doi:10.1371/journal.pone.0181301.

[15] Jeswani, H. K. and Azapagic, A. "Is e-reading environmentally more sustainable than conventional reading?" *Clean Technologies and Environmental Policy* 17 (2015): 803–809, doi:10.1007/s10098-014-0851-3.

How It All Adds Up

[1] Murtaugh, P. A. and Schlax, M. G. "Reproduction and the carbon legacies of individuals." *Global Environmental Change* 19 (2009): 14–20, doi:10.1016/j.gloenvcha.2008.10.007.

[2] Tiefenbeck, V., Staake, T., Roth, K. and Sachs, O. "For better or for worse? Empirical evidence of moral licensing in a behavioral energy conservation campaign." *Energy Policy* 57 (2013): 160–171, doi:10.1016/j.enpol.2013.01.021.

[3] Farmer, J. et al. "Sensitive intervention points in the post-carbon transition." *Science* 364 (2019): 132–134, doi:10.1126/science.aaw7287.

Seth Wynes is studying for a PhD in climate change at the University of British Columbia. He has a Masters in Sustainable Science from Lund University, Sweden, where he co-authored the 2017 research paper, The Climate Mitigation Gap, with Professor Kim Nichols. Prior to studying at Lund University, Seth studied at the University of Western Ontario and McMaster University and taught high school science in England and Northern Canada. Seth's work has been published by the World Resources Institute, the Pacific Institute for Climate Solutions and Environmental Research Letters.

1 3 5 7 9 10 8 6 4 2

Published in 2019 by Ebury Press an imprint of Ebury Publishing, 20 Vauxhall Bridge Road, London SW1V 2SA

Ebury Press is part of the Penguin Random House group of companies whose addresses can be found at global.penguinrandomhouse.com

Penguin Random House UK

First published by Ebury Press in 2019

www.penguin.co.uk

A CIP catalogue record for this book is available from the British Library

ISBN 978 1 529 10589 6

Printed and bound in Great Britain by Clays Ltd, Elcograf S.p.A.

MIX
Paper from responsible sources
FSC
www.fsc.org   FSC® C018179

Penguin Random House is committed to a sustainable future for our business, our readers and our planet. This book is made from Forest Stewardship Council® certified paper.